What Do You Think About Reptiles?

DECIDE FOR YOURSELF

Kalena Baker

What Do You Think About Reptiles?
Copyright © 2025 by Kalena Baker. All rights reserved.
Published by Goose Creek, SC: Teaching Made Practical

No part of this work may be reproduced or transmitted in any form or by any means, electronic or mechanical, including photocopying and recording, or by any information storage or retrieval system, except as may be expressly permitted by the 1976 Copyright Act or in writing from the publisher.

Requests for permission or bulk orders, contact us at www.teachingmadepractical.com/books.
Visit our website at www.teachingmadepractical.com/books

Library of Congress Control Number: 2024921036
ISBN (hardback): 979-8-9915483-1-1
ISBN (paperback): 979-8-9915483-0-4

Editing by Deborah K. Frontiera, *authorsden.com/deborahkfrontiera*
Book design by Monica Thomas for TLC Book Design, *TLCBookDesign.com*
Images by BigStock, *www.bigstockphoto.com* and Adobe Stock, *stock.adobe.com*.

Publisher's Cataloging-in-Publication Data

Names: Baker, Kalena.
Title: What do you think about reptiles? Decide for yourself / Kalena Baker.
Description: Goose Creek, SC : Teaching Made Practical, 2025. | 44 color photos, 2 charts, and 1 map. | Series: What do you think ; book 1. | Audience: Ages 8-11. | Summary: Presents opposing opinions about five different reptiles, each backed by selective facts. After reading the biased information, readers decide their own views on whether alligators are good parents, turtles need shells, snakes are dangerous, lizards are pests, and tortoises make good pets. Questions to encourage critical thinking are included.
Identifiers: LCCN 2024921036 | ISBN 9798991548311 (hardcover) | ISBN 9798991548304 (pbk.)
Subjects: LCSH: Reptiles – Juvenile literature. | Turtles – Juvenile literature. | Snakes – Juvenile literature. | Lizards – Juvenile literature. | Alligators – Juvenile literature. | Critical thinking in children – Juvenile literature. | BISAC: JUVENILE NONFICTION / Animals / Reptiles & Amphibians. | JUVENILE NONFICTION / Animals / Turtles & Tortoises. | JUVENILE NONFICTION / Games & Activities / Questions & Answers.
Classification: LCC QL644.2 B35 2025 | J597.9 B--dc22
LC record available at https://lccn.loc.gov/2024921036

Printed in the United States of America

Table of Contents

About This Book 4
Alligators 6
Turtles 10
Snakes 14
Lizards 18
Tortoises 22
Now It's Your Turn 26
Questions to Think About 28
Glossary 30
Index 31

About This Book

WARNING: This book is a little weird.

Like most reptile books, this book contains interesting reptile facts. However, the facts are presented in an unusual way. Some facts will support one opinion while others will support the opposite viewpoint. Once you've read both sides of the argument, the decision is up to you. Which opinion do you think is right...or do you have a completely different opinion?

As you read, think about how the facts are being used to persuade you to think a certain way. Ask yourself these questions:

- What facts is the author emphasizing?
- What facts is the author ignoring?
- Is the author trying to influence your emotions by making you sad, angry, or disgusted?
- What does the author think is important? Do you agree?
- Does the author's argument make sense?
- Can you trust the author? Why or why not?

Remember, you don't have to agree with everything you read…even if the facts are true!

Alligator Opinion #1

Alligators Are Excellent Parents

Many animal parents leave their babies to fend for themselves. Alligator moms, however, are involved and protective parents.

The excellent parenting starts before their eggs hatch. Alligator moms build nests out of plants and grass—vegetation that helps camouflage the nests. This makes it harder for predators like raccoons to find the eggs.

Alligator hatchlings

As soon as the eggs hatch, mom is ready to help. She gently carries the hatchlings to the water, ensuring her strong jaw does not hurt her babies.

The good parenting does not stop there. Alligator moms fiercely protect their children, keeping them close by for up to 2 years—until they are big enough to scare off predators themselves.

Alligator moms protect their children until they are big enough to protect themselves.

Alligator Opinion #2

Alligators Are Terrible Parents

While alligator moms attempt to protect their young, they often fail. Many of their eggs will never even hatch! Some of the eggs will be destroyed by water. Others will be eaten by raccoons, skunks, or other predators. Some careless alligator moms even crush their own eggs!

Alligator dads are even worse. They do not make **any** attempt to protect their babies. In fact, scientists believe that alligator dads do not recognize their own children! Sometimes, adult male alligators will snack on a hatchling for dinner—possibly eating their own child without realizing it.

? What do you think? Are alligators good parents?

Baby alligator

Baby alligators face danger from many predators—raccoons, blue herons, otters, and even adult alligators!

Turtle Opinion #1

Turtles Need Their Shells

Without a shell, it would be almost impossible for turtles to protect themselves. Shells are like armor for a turtle, safeguarding body parts and organs.

Turtles do not move fast, but that is ok. The protection a shell provides makes up for the slow speed. Turtles do not have to run away when a predator gets too close. Instead, they can hide in their shell and wait for their enemies to leave! The odd shape and strength of the shell make it difficult for other animals like foxes, cats, and snakes to bite. Predators will often give up in search of prey that is easier to eat.

Turtle shells also provide camouflage, further protecting them from predators.

Turtle Opinion #2

Turtles Don't Need Their Shells

Turtles are notoriously slow—and their shells are to blame! Shells are heavy and require extra energy to move. They are also rigid and inflexible, limiting range of motion and making turtles clumsy.

If the shell offered better protection, it might be worth the slow speed. But animals like alligators and honey badgers can easily bite through a turtle's shell. Large birds are also able to break turtle shells. They will fly small turtles high into the air and then drop them, cracking their shell and making them easy prey.

Without a shell, perhaps turtles could outrun some of their bigger predators!

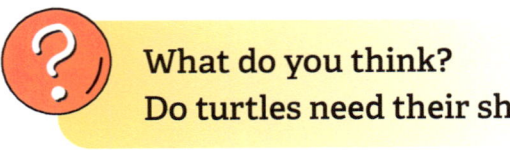

What do you think?
Do turtles need their shells?

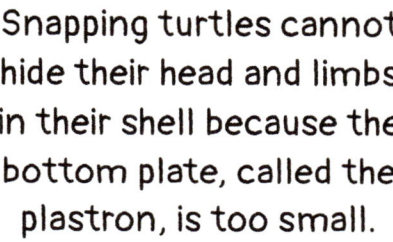

Snapping turtles cannot hide their head and limbs in their shell because the bottom plate, called the plastron, is too small.

Sea turtles cannot hide in their shells, which makes them easy prey when they are nesting on land.

Snake Opinion #1

Snakes are Not Dangerous

Encountering a snake can be terrifying, and people often react in fear. But are snakes really that dangerous?

Snakes prefer hiding over attacking, although they might defend themselves if provoked. Most are nonvenomous, so they cannot do much damage anyway.

In the United States, an average of 5 people die from a snake bite each year. Bees, hornets, and wasps, in contrast, are fourteen times more lethal, killing approximately 72 Americans every year.

Average Number of Deaths Per Year in the U.S.	
Snakes	5
Lightning	28
Dogs	43
Hornets, wasps, and bees	72
Smoking	480,000

Data from Centers for Disease Control and Prevention

Boas are not venomous.

Snake Opinion #2

Snakes are Dangerous

Many people are frightened of these legless, scaly reptiles—and for good reason! Approximately 100,000 people around the world die every year because of snakes—and millions more have lifelong struggles because of an interaction with a snake.

While most snakes are nonvenomous, the venomous ones are extremely dangerous. For example, a king cobra's venom has the strength to take down an adult elephant in just a few hours.

In general, snakes will not attack unless they feel threatened. Unfortunately, it is easy to stand near a well-camouflaged snake without even realizing it, unintentionally threatening the snake.

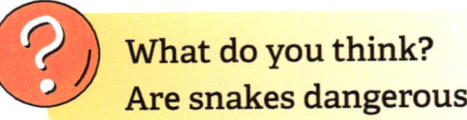

What do you think? Are snakes dangerous?

How Many Snake Bites Around The World...	
Happen each year?	4.5–5.4 million
Cause longterm health problems each year?	1.8–2.7 million
Lead to death each year?	81,000–138,000

Adapted from https://www.who.int/news-room/fact-sheets/detail/snakebite-envenoming, accessed 7/17/24. WHO is not responsible for the content or accuracy of this adaptation.

Boas are not venomous. Instead, they kill their prey by wrapping around it tightly, cutting off blood circulation.

Lizard Opinion #1

Lizards are Pests

Sure, lizards are cute—but when they live near humans, they become pests. Lizards are sneaky and slyly invade homes, restaurants, and offices.

Lizards can cause all sorts of damage when they are inside buildings. They have been known to mess with electrical wiring, which is expensive to replace. They often carry salmonella in their digestive tracts and can spread this disease to humans. And who wants to deal with lizard poop in the kitchen?

Also, lizard bites can be painful. Some lizards are even venomous. How can a venomous creature **not** be a nuisance to humans?

> **Venomous Lizards**
> - Gila monster
> - Komodo dragon
> - Mexican beaded lizard
>
> (pictured below)

In Florida, Georgia, and South Carolina, the Tegu lizard is an invasive species. It does not belong and is bad for the animals and plants that live there.

Lizard Opinion #2

Lizards are Helpful

Have you ever been annoyed by the bugs in your house? Then you need some lizards!

These reptiles **love** to eat insects. Lizards help keep insect populations under control, limiting the number of mosquitoes, ants, termites, and cockroaches running around. Given the choice, would you rather encounter a lizard or a cockroach?

Also, lizards are harmless. They might bite if threatened, but these bites do little damage. A few species are venomous, but it is rare to come across one.

Some people value lizards so much that they keep them as pets!

What do you think? Are lizards pests?

Komodo dragons are venomous, but you will most likely never see one. They only live in Indonesia!

Tortoise Opinion #1

Tortoises Make Terrible Pets

Pet tortoises are a **huge** responsibility. Setting up an enclosure where tortoises can thrive takes careful planning. Not only do they need plenty of space, but they also need special UVB lighting to keep their bones healthy. The temperature and humidity inside the enclosure must be carefully monitored as well.

After the enclosure is ready, be prepared to financially provide for your tortoise for a looooooong time. Some tortoises can live to be 80 years old! Tortoises are a life-long, expensive obligation, making them terrible pets.

Did you know that all tortoises are turtles, but not all turtles are tortoises? Tortoises live on land and have stumpy legs and heavy shells.

Some tortoises do not like being handled by humans. It can stress them out!

Tortoise Opinion #2

Tortoises Make Fantastic Pets

Dogs shed. Cats scratch. Hamsters only live a few years. But tortoises do not have any of these downsides, making them excellent pets!

Once tortoise enclosures have been set up, they just need food and love. These reptiles are quiet, harmless, and can thrive without much attention.

Tortoises are interesting animals, each with their own personality. Since they live longer than most pets, tortoise owners have plenty of time to get to know and enjoy their pets!

 What do you think? Do tortoises make good pets?

Tortoises are gentle and safe reptiles!

Now It's Your Turn

Use the crocodile facts below to write 2 different opinions:

1. Crocodiles should be removed from their homes to protect humans.
2. Crocodiles should not be removed from their homes.

Crocodile Facts

- Saltwater crocodiles have the strongest bite of any animal in the world.
- 7 species of crocodiles have a high risk of going extinct.
- Freshwater crocodiles rarely attack humans, instead avoiding them.
- Some crocodiles are more than 20 feet long and weigh more than 2,000 pounds.
- Crocodiles that have been relocated to new places often return to their homes.
- Saltwater crocodiles kill around 1,000 people a year.

 **What do you think? Should crocodiles be removed from their homes?**

Most crocodile attacks occur in the water and areas that humans can avoid.

Questions to Think About

Alligators
- What does the first alligator passage say about alligator dads? What does that tell you about the author?
- Do you think it is fair to expect an alligator mom to keep all of her eggs alive?

Turtles
- The first turtle passage says a turtle's shell is hard for other animals to bite, but the second passage says that some animals can easily break it. How do you think both statements can be true?
  - Describe how each turtle passage thinks turtles can best protect themselves from predators. What do you think works best to protect them?

Snakes
- The first snake chart says that 5 people die every year from snake bites, while the other chart says that around 81,000–138,000 die every year. How can both be true? Which chart is more helpful for you?
- Compare the two pictures of boas. How are the pictures and captions similar, and how are they different?

 # Questions to Think About

Lizards

- How did you feel when the first lizard passage mentioned lizard poop? How did the author intentionally stir up emotion in you?

- The second lizard passage asks whether you would rather see a lizard or a cockroach. Is this a fair question? Do you have to pick between these two?

Tortoises

- According to the first tortoise passage, what makes a good pet? What makes a good pet according to the second passage? What do you think makes a good pet?

- How does each tortoise passage talk about setting up a tortoise enclosure? What information is missing from the second passage? Do you think one author exaggerates how much it would cost?

Remember, you do not have to agree with every opinion you read—even if the facts are true!

Glossary

Camouflage—when animals use colors or patterns to blend into their surroundings so they cannot be seen easily

Enclosure—an area that has walls or a fence that helps keep animals or things inside

Fact—a statement that can be proven true

Invasive species—a plant or animal that moves to a new place and causes problems for the plants and animals that already live there

Opinion—a statement that tells a personal feeling or belief

Pest—something that causes problems or bugs you

Predator—an animal that hunts and eats other animals

Prey—an animal that gets hunted and eaten by another animal

Reptile—cold-blooded animals that breath air, have scales, and lay eggs

Salmonella—a germ that can give you a fever and stomach pains

Venomous—an animal that has a poison inside of it that it uses to defend itself through a bite or sting

31

Index

Alligator, 6–9, 12, 28

Camouflage, 6, 11, 16, 30

Crocodile, 26–27

Eggs, 6, 8, 28, 30

Lizard, 18–21, 29

Predator, 6, 8–12, 28, 30

Prey, 10, 12–13, 17, 30

Snake, 10, 14–17, 28

Tortoise, 22–25, 29

Turtle, 10–13, 22, 28

Venomous, 14–18, 20–21, 30

Parents & Teachers

Get free printables to use with this book at www.teachingmadepractical.com/whatdoyouthinkaboutreptiles.

www.ingramcontent.com/pod-product-compliance
Lightning Source LLC
Chambersburg PA
CBHW051515110526
44582CB00007B/131